The San Diego-La Jolla Underwater Park Ecological Reserve

La Jolla Shores & Canyon

Judith Lea Garfield

Picaro Publishing - La Jolla, California

Published by Picaro Publishing, P.O. Box 332 La Jolla, CA 92038-0332
© 2000 by Judith Lea Garfield

First edition published in 2000.
Printed in Singapore.

06 05 04 03 02 01 00 6 5 4 3 2 1

Photo credits: All photos by the author except as noted;
Steve Gardner, pp.23, 36, 56 (F); Tracy Clark, p.56 (M) ; Phil Colla, p.59

Garfield, Judith Lea.
The San Diego-La Jolla Underwater Park Ecological Reserve:
La Jolla Shores and Canyon.

Library of Congress Control Number: 00-090373
ISBN 0-9640724-7-5

Comments?   **Additional Orders?**   **Questions?**

Write to: Judith Garfield
              c/o Picaro Publishing
              P.O. Box 332 La Jolla, CA 92038-0332
Email: judith@garfield.org
Visit: www.judith.garfield.org

Front Cover: La Jolla Shores, La Jolla, California

To Sydney and Sasha, my best friends.

To my mom, Betty,
for her love and enduring support.

To everyone who recognizes
the value of our ocean
and acts to protect its legacy.

Foreword

Judith Garfield has done it again! In Volume 1 of this series, Judith set the standard for divers who dive in rocky habitats in shallow San Diego waters, especially in the La Jolla area. That volume has been the only field guide that is needed to rapidly identify marine species, grasp quick information on their behavior, and even outline the diving areas of interest. Now this Volume 2 gives readers the wealth of information and references that cover the sandy habitats, especially in the La Jolla Submarine Canyon. Together with Volume 1, this newest handy field guide will give the San Diego (and even the Southern California) diver all that is needed to gain the knowledge and familiarity of most of the major marine environments in San Diego.

Thanks to Judith's determination, stamina, and marvelous observation powers, Volume 2 shapes up as the reference on sand life in the San Diego-La Jolla Underwater Park and Ecological Reserve. She has presented a fabric of species that commonly inhabit the edge of the La Jolla Canyon, and her camera has captured these species in all their splendor. Just as in Volume 1, this current work is magnificent in its artistry, complete and accurate in details, and follows the logical sequence of evolutionary trends in its organization.

As an added bonus, Judith has finally come up with names to all of the dive sites around La Jolla Canyon, and steady use of Volume 2 will allow all divers to pinpoint dive sites. Instead of diving La Jolla Canyon, divers in the future will meet at Vallecitos Street for a dive to East Wall or park close to the La Jolla Beach and Tennis Club for a dive to Pipefish Patch.

I envision that once divers and snorkelers discover that Volume 2 is available, the two volumes will be standard reference for marine life and dive sites in the La Jolla area. Judith is to be commended for her persistence in authoring these two volumes.

This useful field guide is a must for any novice or even experienced diver who wishes to learn more about marine life in the La Jolla Canyon waters without taking a marine biology course. I highly recommend carrying this handy guide to browse though both prior to and after a dive to a sandy habitat.

Bert N. Kobayashi

Dr. Bert N. Kobayashi
Marine Biologist
Platinum Pro 5000 Diver and Instructor
University of California, San Diego

Contents

Acknowledgments

I have so many people to thank for making this book possible. On a personal level, I am indebted to my great friend and dive buddy, Christine Schumann, for so much, especially tolerating our repetitive scans over the canyon; Pat Schaelchlin, for her friendship and encouragement, as well as her knowledge of La Jolla; Diving Unlimited International, especially Pam Oliva, who kept me diving dry and did so without keeping me dry-docked for long; and Elizabeth Whalen, wonderful friend, knowledgeable professional, and outstanding teacher.

On a professional level, I thank Dr. Bert Kobayashi for, once again, invaluable technical editing and writing the Foreword; Steve Gardner, Phil Colla, and Tracy Clark for important photographic contributions; Kai Schumann for photo assistance in creating the back cover photo; JoAnn Padgett for outstanding copyediting; Blaise Nauyokas for yet another knockout layout; Troy Murphree for developing the inside cover chart-map, Peter Langenfeld for making the final chart-map look professional, and George Silva for the final map flourishes.

Of course, this book would not be possible without the financial help of all the divers and other ocean enthusiasts who trusted me to make a another high-quality book that would add to their appreciation and knowledge of our underwater park. In this regard, I give extra thanks to Ocean Enterprises and especially Terry Nicklin and the Diving Locker crew.

Introduction

The ecological reserve comprises four different underwater landscapes: rocky reefs, sandy flats, a kelp bed, and a submarine canyon. Vol.1 in this series about the underwater park showcases the most common marine life found in the ecological reserve's shallow (30 feet deep or less) waters off La Jolla Cove but emphasizes only a portion of the reserve's varied terrain – the rocky reefs and kelp bed.

Vol. 2 journeys into the reserve off the nearby waters of La Jolla Shores to explore the other two habitats, the vast sandy flats and submarine canyon, and identify the very different kinds of marine life that thrive there.

Keep in mind that this book is not meant to cover everything you can see in the reserve. Rather, it presents a variety of the most common species. As with Vol. 1, you will find that even the most common subjects lead the most uncommon lives; therefore, an emphasis is placed on unusual behaviors, lifestyles, and other habits. Knowledge of these distinguishing characteristics is particularly valuable to divers because when you know what to look for, you dramatically increase your chance of discovery.

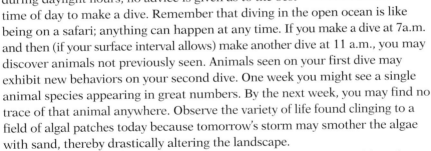

Sites listed in the Points of Interest are arranged alphabetically. However, marine life species are organized according to the standard evolutionary scale. For example, green algae are considered more primitive than red; sea slugs are more primitive than octopuses; and sharks are the most primitive of all fishes listed.

While this book is a record of the species found during daylight hours, no advice is given as to the best time of day to make a dive. Remember that diving in the open ocean is like being on a safari; anything can happen at any time. If you make a dive at 7a.m. and then (if your surface interval allows) make another dive at 11 a.m., you may discover animals not previously seen. Animals seen on your first dive may exhibit new behaviors on your second dive. One week you might see a single animal species appearing in great numbers. By the next week, you may find no trace of that animal anywhere. Observe the variety of life found clinging to a field of algal patches today because tomorrow's storm may smother the algae with sand, thereby drastically altering the landscape.

Wreckage or other underwater structures are not discussed. Although it is illegal to place or dump any object into the park, tidal flow prevents any matter from remaining in the area for any period of time; the forces of water carry any loose objects to the canyon's edge where they ultimately roll over into the abyss.

Whether you are a landlubber or a diver, Vol. 2 (along with Vol. 1), will provide a clear picture of the ecological reserve's history, geology, topography, and surroundings, as well as knowledge about the plants and animals living along the Southern California coast.

Use these books as stepping-off guides for your own personal exploration and discovery. Jump in, expect the unexpected, and don't forget your dive light and compass!

The Ecological Reserve

In 1970, the San Diego City Council dedicated 6,000 acres of submerged lands – from San Diego's northern border at Torrey Pines State Park to La Jolla Cove – to be created as an underwater park. The City of San Diego's Department of Parks and Recreation and the California Department of Fish and Game would share responsibility for park maintenance and enforcement.

The following year, the Council created the ecological reserve within the park as a "look but don't touch" area. The reserve encompasses just over 1.5 miles of coastline that includes La Jolla Shores and La Jolla Canyon. The reserve's coastline boundary stretches from the University of California's Scripps

Institution of Oceanography southern border at La Jolla Shores, continues south, then bends west around Devil's Slide all the way to La Jolla Cove. The boundary across open water is marked by large yellow buoys and extends out to the deepest part of La Jolla Canyon (about 1,000 feet deep).

The ecological reserve is invaluable because it affords protection to marine life dwelling within four landscapes. Off the California coast, you can find rocky reefs, sandy flats, kelp beds, and submarine canyons. Most regions harbor one or maybe two of these landscapes within a small area, but the ecological reserve comprises all four. What does this mean? Consider a place that embodies the Grand Canyon, a redwood forest, a desert, and any place that's great for boulder-climbing, and you have the ecological reserve. Divers also benefit from the reserve since the landscapes embrace such diverse marine life.

By creating a reserve, we have acquired a nursery of sorts – a safe breeding ground for the huge number of diverse life forms found off our coast. Formation of the reserve has been critical in reestablishing depleted marine life species, safeguarding the area's fragile ecology, and preserving the natural beauty of the shoreline. While park officials have been somewhat successful in providing protection for the flora and fauna, community education must continue to prevent ongoing degradation from poaching and littering.

Rules and Regulations
(Enacted March 8, 1971)

Pursuant to Section 1580 through 1584 of Fish and Game Code, the San Diego-La Jolla Ecological Reserve is to be preserved in a natural condition for the benefit of the general public to observe native flora and fauna:

"No person shall disturb or take any plant, bird, mammal, fish, mollusc, crustacean, reptile or any other form of plant life, marine life, geological formation, or archaeological artifacts..."

Enforced by the State of California Department of Fish and Game
Anyone who violates this code will be subject to a minimum of $1,000 in fines, imprisonment in county jail for six months, or both. If you see illegal activity, call the 24-hour number: **(888) DFG-CAL-TIP.**

Topography and Geology

In 1907, scientists at Scripps Institution of Oceanography discovered a submarine canyon after taking soundings from their boat off La Jolla Shores. Plunging to about 1,000 feet deep, the explorers named the canyon Soledad Drowned Valley. Eventually, it came to be known as La Jolla Canyon. It is San Diego's most popular scuba diving area for its variety of terrain and marine life and because the waves that splash onto its mile-long, sandy beach are usually the most gentle of all San Diego beaches.

To take your own tour of La Jolla Canyon, enter the water at La Jolla Shores. As you swim west through shallow waters (10 to 25 feet) over the flat bottom, you pass over sand made up of fine-grained quartz and intermeshed with feldspar, hornblend, and mica. After you swim out 300 to 400 yards (depending upon where you entered the water), the sandy bottom drifts downward to a depth of 30 to 40 feet. Suddenly, you experience a sharp drop-off that signals your entrance into the canyon.

The canyon is shaped like a wide bowl with the slopes of the bowl made up of a soft, gooey, peaty-clay material. Whether you kick to your right or left, stay within depths of about 40 to 85 feet, because this is where you will find the numerous terraced ledges that harbor the majority of canyon marine life. The upper ledge, or lip, is made up of a crumbly sandstone-shale material. As you navigate, do not expect to maintain a steady compass course, because the canyon is not evenly carved; you will meander in and out as you periodically encounter gently sloping areas, gullies, steep cliffs, or wide valleys.

In parts of the canyon where cliffs exist, you may feel like a mountain climber as you scale the sides of the canyon wall (with fins on!). In some areas, the faces of these walls become so steep that they are nearly vertical and up to 20 feet high. The walls are the areas of richest marine life.

Continue along the well-defined cliffs; they ultimately become shallower and shallower until they disappear altogether. You now find yourself in a valley of sand. These valleys are often carpeted with masses of decaying marine plants that look like large, dark mats resting on the sand. You may even note some gas bubbles (maybe methane) escaping from this compost, and taste sulfur in the water.

As you make your way back to shore, you may think that the desertlike landscape of the canyon and sandy flats makes for a bleak and empty place. But the moonlike austerity is merely a foil for the number and variety of colorful marine life that live there. In fact, it is this subtlety that makes addicts of so many divers who will never have enough time to explore what lives on and in the vast expanse of rippled sandy bottom and who will never tire of the canyon's varied landscape, the perpetual night of its depths, and the unfolding dramas of its living inhabitants.

An Underwater Archaeological Site

In 1954, an important archaeological site was discovered off La Jolla Shores, just offshore from the La Jolla Beach and Tennis Club. Using a new form of submarine exploration – scuba diving – divers discovered and hauled out thousands of stone mortars. Grinding stones, fine scrapers, and pounding stones were also recovered.

These stone tools were the remnants of a prehistoric society. But why was the site under water? Who were these people? How did they live? What happened to them?

To begin to answer these questions, we must go back to the last Ice Age (over 17,000 years ago) when glaciers covered much of the earth and, in turn, comprised most of the earth's water. At this time, our ocean was a much smaller body of water. Even when the Ice Age ended 17,000 years ago and the sea began to rise, more coastline was exposed than we see today. In fact, had you lived along La Jolla Shores back then, you could have walked about 275 yards from La Jolla Shores out to the rim of La Jolla Canyon. (It is now under 60 feet of water!)

The early dwellers of the La Jolla Shores site lived from 9,000 to 3,000 years ago and were made up of a group or groups of people that anthropologists named the La Jollans. Not much is known definitively about their lives; however, many archaeologists think that they lived off La Jolla Shores (on land that is now under water) around a freshwater lagoon that was cut off from the sea by a sandy barrier island that paralleled the shoreline.

The La Jollans may have survived because the then damper climate brought lots of rain, which kept the many nearshore saltwater lagoons filled to over-flowing. When the spillover flooded back into the ocean, fish and other marine life washed into these shallow, natural fish ponds. In this way, the natives were supplied with a steady stock of easily available food.

After 14,000 years (3,000 years ago) of melting glaciers, the rising water level finally drowned the lagoon. During this same time period, the earth's weather changed dramatically – temperatures increased, rainfall decreased, other nearby lagoons started silting, and marine and land food supplies began to dry up. Eventually, the La Jollans were forced to abandon their oceanfront site for places unknown and, ultimately, they faded into history.

Today, the site in which the early La Jollans lived is a watery grave. As tidal flow and storms shift the sands, other parts of this site may reveal themselves intact. While those unable to don a mask and regulator cannot visit here, scuba divers must remember to admire any encountered artifacts with their eyes only. By law, all artifacts are protected (See p.8, Rules and Regulations).

Let us respect the reserve's rules and regulations so that for years to come, visitors may thrill to discover – or rediscover – the treasures left by these ancient peoples.

Points of Interest

LA JOLLA SHORES SANDY FLATS*

Eel Grass Islands Oval patches of emerald green "lawn" appear like oases in a sandy desert. Lobsters (p.38) hide in the tall grass, small animals cling to the blades (pp.31,36,50), and juvenile fish (Vol. 1, pp.40,47,48,51) cruise above. *Access from: Marine Room restaurant; Topside heading: 270°; Shore distance: 1,000 ft.; Water depth: 30-50 ft.*

Pipefish Patch Furry coverings on pebbles and rocks are actually a mix of different algae that sprout along this cobbled area. Look closely to discover the well-camouflaged pipefish (p.50) sporting a mix of patterns and colors to better blend into the tufts of red, green, and brown plants, particularly thin dragon beard algae (p.12). Crustaceans (pp.36,39) and moon snails (p.21) reside here. *Access from: Marine Room or Vallecitos Street; Topside heading: between Sea Lodge hotel and Marine Room restaurant; Shore distance: 400 ft.; Water depth: 7-25 ft.*

Shark City From the surf zone and just beyond, leopard sharks (p.44) and shovelnose guitarfish (p.45) slither along the sandy bottom; they are most abundant during summer when they arrive in numbers to mate. Observe them by snorkeling or floating on the surface, as tank bubbles terrify these shy creatures. Look for bat rays (p.49), round stingrays (p.47), and an occasional butterfly ray (p.48). *Access from: Marine Room; Topside heading: in front of Marine Room restaurant; Shore distance: surf zone and just beyond; Water depth: 3-8 ft.*

LA JOLLA CANYON

Gorgonian Gardens Head down to 87 feet; the red gorgonian fans (p.18) look brown at this depth, so use your dive light to illuminate their true color. Continue north, dipping down to 100 feet, to see more of these small treelike animals. Look for sarcastic fringeheads (p.57) and tube-dwelling anemones (p.17) often found nestled near the base of the gorgonians. *Access from: Vallecitos Street; Topside heading: 305°; Shore distance: 1,000 ft; Water depth: 87-100 ft. (Plan this deep dive carefully, and dive within the decompression tables.)*

North Wall Point Continue north or south on this deep, steep wall that is lush with life. On sandy bottom, just above the top of the ledge, pass by a grove of sea pens (p.16) and glum-looking spotted sand bass (p.55). Brown rockfish (p.52) rest vertically in narrow crevices, clusters of red rock shrimp (p.37) align themselves in sandstone cracks, and sea slugs (pp.26-31) adorn the wall with splotches of brilliant color. *Access from: south end of parking lot; Topside heading: 305°; Shore distance: 1,700 ft.; Water depth: 60-85 ft.*

Vallecitos Point Head north as the ledge steepens around this promontory. Along the upper ledge, tiny brilliantly colored blue-banded goby fish (p.29) stake out their territory, and cabezon (p.54) guard their eggs; along the lower ledge, look for numerous large excavated holes. Find lobster (p.38), sheephead (p.56), brown rockfish (p.52), scorpionfish (p.51), and red rock shrimp (p.37) in these dens. *Access from: Vallecitos Street; Topside heading: 300°; Shore distance: 1,000 ft.; Water depth: 50-70 ft.*

* Because various water conditions (swell, surf, currents, and tidal flow) cause the terrain to shift in these shallow sandy areas, expect dramatic changes of scenery (e.g., marine life appearance or disappearance) from season to season or year to year.

Appearance: Slender, delicate, and deep reddish-brown; irregular and meager erect cylindrical branches arise from a central "stem."

Size: From 4 to 18 inches in height.

Where Found: Along La Jolla Shores attached to sand-covered rocks from 15-30 feet deep (e.g., Pipefish Patch).

Feeds On: Sunlight and dissolved matter (photosynthesizes).

Thin Dragon Beard
Gracilaria sjoestedtii

One of the most common red algae in Southern California intertidal regions, thin dragon beard is found from British Columbia to Costa Rica, as well as in Peru and China. While it has a rather stiff texture, young plants can be eaten. Because of its remarkable powers of regeneration, this red alga is easily cultivated artificially; plants are simply chopped into small pieces and hung in net containers about three feet below the surface where they grow into many new plants. Dragon beard is harvested commercially in China and made into agar (a type of gelatin). Look for the bay pipefish (p.50) and skeleton shrimp (p.36) that often hide among the wispy strands.

Flower pod with seeds

Eel Grass

Zostera marina

In the same category as surf grass (Vol.1, p.15), eel grass is not a primitive seaweed but a true rooted flowering plant much like a plant on land. Eel grass flowers bloom as yellow-green flower pods throughout the summer months. Since eel grass self-pollinates (it contains both male and female sexual structures), the flowers are not showy displays. Each flower pod contains 10 to 30 seeds. When ready, the pod splits and releases dense seeds into the water where they sink down and root. Look for skeleton shrimp (p.36), tiny snails, California spiny lobsters (p.38), crabs, and fishes (Vol.1, p.40, 47, 58, 64) around the eel grass oases.

Appearance: Dense bundles of long, green, grasslike blades anchor in soil via fine roots; flowers are produced on stalks.

Size: Blades to 3 feet in height. Flower stalks to 4 feet in height.

Where Found: Sheltered sandy bottom off the Marine Room toward South Wall near the canyon rim from 30-55 feet deep (Eel Grass Islands).

Feeds On: Sunlight and dissolved matter (photosynthesizes).

Strawberry Anemone
Corynactis californica

Appearance: Red, pink, orange, tan, yellow, or buff-colored stalks that flare out toward the crown; crown made up of white tentacles.

Size: To about 1/2 inch in height and diameter for each individual stalk. The crown spans up to 1 inch in diameter.

Where Found: Spotty appearance throughout the canyon, along the ledges, and attached to rocks from 50-100 feet deep.

Feeds On: Tiny animals drifting by.

While in the same group as the wandering jellyfish, this homebody anemone is content to stay anchored to the bottom. Every one of the anemone's tentacled tips harbors tiny balloons of stinging cells. Because each balloon is under tremendous pressure (140 atmospheres or the equivalent pressure of diving in water over 4,000 feet deep), the balloons burst at the whisper of a touch by any living thing. The small prey that blunder within the anemone's tentacled reach burst the poison balloons, become paralyzed, and are eaten. Humans are not affected by the tentacle's toxins.

Sea Pansy

Renilla kollikeri

The sea pansy, a colonial animal (many individuals living together as one), captures food by spinning a mucus web over its surface, then traps drifting animals in this net. The white tentacles of each polyp then sting the victims, and the net – plus its contents – are sucked down into the pansy's polyp mouths and swallowed. Although it lacks arms and legs, a pansy is not immobile. It creeps along the bottom by contracting muscles located along the edges of its petal. It is defenseless against its predator, the armored sea star (p.41), as the sea star devours the entire pansy.

Appearance: A flat, heart-shaped, purple disc (body) is spread out like a pansy petal. Disc top is usually dusted with a light layer of sand from which feeding polyps (tiny anemonelike bodies) with white tentacles protrude. A fleshy stem (penduncle) anchors pansy in the sand.

Size: To about 3 inches in diameter.

Where Found: In all sandy areas from 10-50 feet deep.

Feeds On: Microscopic animals floating by.

Appearance: Central stem is
white, stiff, and brittle.
Lateral branches of fluffy
white, tan, or gray plumes
radiate from the stem.
Overall, presents a quill-
like appearance.

Size: Usually to about 1 foot
in height.

Where Found: La Jolla Shores,
especially abundant along
canyon's East Wall, from
10-100 feet deep.

Feeds On: Microscopic plants
and animals.

Slender Sea Pen
Stylatula elongata

Looking more like a 19th century writing
tool than an animal, the sea pen (a colonial
animal like the sea pansy, p.15) displays
either a very slim or more full-bodied
appearance. It can do so because of its
plumes, which are made up of tiny individual
anemonelike bodies with mouths that can
drink in large quantities of water. In this
way, the sea pen presents a lusher appearance
and, in doing so, gains a larger area to sift
through; thus, it captures more prey drifting
by. While the sea pen has a bulblike structure
that allows it to anchor in one place in the
mud, it can quickly sink down into its sand
burrow when danger signals the need for a
quick getaway.

Tube-Dwelling Anemone

Pachycerianthus fimbratus

The tube-dwelling anemone mimics the same flowery look as other anemones (p.14; Vol.1, p.17). This anemone's tormentor is the nudibranch, giant Dendronotus (p.29). This sea slug dines with gusto on the anemone's spaghettilike tentacles. When the nudibranch attacks a tentacle strand, the anemone withdraws into its tube, often taking the unconcerned sea slug with it. The nudibranch simply finishes its meal and crawls out the tube. Apparently, the anemone is none the worse for wear from these assaults and is left only with a bad haircut. The tube-dwelling anemone may live for 10 years or more.

Appearance: Long, slim, whiplike, tentacles are cream, tan, or orange. Tube is tough, slippery, and tan or black. When disturbed, tentacles close down completely into the tube.

Size: Tube reaches about 4 inches in height and 1 inch in diameter.

Where Found: In the canyon sitting alone on soft, muddy hillsides from 50-100 feet deep. See them in Gorgonian Gardens embedded in mud directly next to a red gorgonian (p.18).

Feeds On: Small animals drifting by.

Appearance: Tough skeleton is deep red and treelike with main and side branches formed in a single plane. Branches are covered with polyps; each polyp wears a crown of furry, white tentacles.

Size: To 2 feet in height.

Where Found: In Gorgonian Gardens from 80-100 feet deep. Elsewhere in the canyon from 85 feet deep.

Feeds On: Microscopic plants and animals.

Red Gorgonian
Lophogorgia chilensis

In the dimly lit sea, the gorgonian (also called a sea fan) looks more like the charred remnant of a fire's aftermath; however, shine a light on this seemingly dead, brown twig, and the fan's true color is instantly revealed. A flower animal like the anemone (pp.14,17; Vol. 1 p.17), the gorgonian is not a single animal, but hundreds of tiny, tentacled sea anemone-like creatures called polyps. The polyps cooperate to build a large fan structure out of hornlike material; they align their tree crosswise to the current to let each polyp capture the maximum amount of floating food. Look for simnea (p.22), a tiny mollusc that lives and feeds exclusively on this fan.

Lacy Bryozoan

Phidolopora pacifica

Almost entirely overlooked by divers, the lacy bryozoan is as fragile as a potato chip. A colonial animal, each bryozoan mass is a maze of hollow chambers that are home to many individuals. Each animal wears a crown of tentacles that it waves in the water to capture food. The tentacles are full of stinging bodies; they paralyze the prey, then steer the captive into an individual bryozoan's mouth. Should it sense danger, each bryozoan individual can pull its tentacles back down into the safety of its calcium-constructed condominium.

Appearance: Patches of delicate, fluted, orange to orange-brown lace form a convoluted, low profile mass.

Size: Mass from 4 to 7 inches across and from 2 to 4^1/$_2$ inches high.

Where Found: Attached to ledge outcroppings throughout the canyon from 50-80 feet deep.

Feeds On: Microscopic plants and animals.

Trough-Nose Worm

Echiura sp. (unidentified)

Appearance: Long, flexible, trough-shaped, appendage (proboscis) extends out its circular burrow and lies against the vertical face of the ledge.

Size: Proboscis to about 4 inches in length and 1/8 inch in width.

Where Found: Burrowed into the canyon's ledge, with only its proboscis exposed, from 55-75 feet deep.

Feeds On: Decomposed plants and animals.

This worm has a sausage-shaped body that stays firmly hidden within its burrow. Its only visible part is its proboscis (noselike structure), a muscular organ, capable of tremendous extension, that is unique to any other in the animal kingdom. To feed, the worm extends its trough-shaped proboscis out the burrow and scoops up the top layer of dead material that accumulates on the mud's surface. The proboscis mixes this food material with mucus, then carries the stringy concoction back to its mouth where it is swallowed. When you see the trough-nose, maintain some distance, as sudden water movement causes the worm to quickly retract its "nose" into its burrow.

Sand collar egg mass

Lewis' Moon Snail

Polinices lewisii

Usually, the moon snail is seen plowing through the sand almost completely submerged. Its monstrous foot does not look like it can fit completely into its shell – but it will. When threatened, the snail sprays water like a garden sprinkler out tiny openings on its foot; in seconds, all the water is squeezed out, and the formerly huge foot is now small enough to pull into its shell. Moon snail eggs appear as a broad, curved sand collar. Shaped around the adult's shell, the stiff, rubbery tan collar (resembling a broken gasket) is made up of one layer of tiny eggs sandwiched between two layers of sand; the mass is held together by mucus.

Appearance: Smooth, round, thick, yellow-brown shell. Body (mantle) and foot are pale-gray to white. Two white horns outlined dark-brown protrude just below the front of the shell.

Size: Shell to 3 inches across; extended foot much larger than shell. "Sand collar" egg mass to 5 inches in outside diameter.

Where Found: Along the shallow sandy bottom to 30 feet deep.

Feeds On: Adults – clams; juveniles – algae.

Appearance: Smooth, rose-colored shell; body is pink with red and yellow dots.

Size: To about 1 inch in length.

Where Found: On its host, the red gorgonian (p.18), from 70-100 feet deep.

Feeds On: Red gorgonians.

Simnea

Delonovolva aequalis

This mollusc is closely related to a true cowrie, such as the chestnut cowrie (p.23). Worldwide in distribution, most of these species are tropical. The simnea is a carnivore that lives and feeds on polyps (tiny anemonelike bodies) that make up and live in the gorgonian structure. The simnea's shell, body, and patterns match the color of the fan it feeds on. Because it lives on such a specialized and restricted habitat, it has evolved in a competitive way seldom seen in other molluscs. The male controls a major part of the fan on which he lives and will ward off other males that encroach on his territory.

Chestnut Cowrie

Cypraea spadicea

Since prehistoric times, people have used the shiny porcelainlike cowrie shell both as an ornament and for money. Unlike most other encrusted seashells, no plants or animals settle on the cowrie shell. How does it keep hitchhikers at bay? When active, the cowrie stretches a thin layer of its tissue over its entire shell to create a slippery barrier to potential settlers. Most of the approximately 185 species of cowries are tropical; only the chestnut cowrie is common in California's temperate waters.

Appearance: Highly polished shell has a light chestnut-brown patch surrounded by a white border; body is apricot-tan with darker spots. If undisturbed, the body slides up to cover the shell.

Size: To 2 1/2 inches in length.

Where Found: Along crevices and ledges mostly along the canyon's North Wall from 60-80 feet deep.

Feeds On: Sponges, small sea anemones, and snail eggs.

Purple Olive
Olivella biplicata

Appearance: Olive-shaped, smooth, shiny shell is lavender, white, grayish-black, or yellow. Foot is white or cream-colored.

Size: Usually no more than 3/4 inch in length.

Where Found: La Jolla Shores just past the surf zone to 15 feet deep, often in clusters of 25 or more. Look for a raised, single plowed track in the sand to reveal this underground traveler.

Feeds On: Plants and dead animals.

The burrowing purple olive has a creative way of breathing while tunneling under the sand. It uses its tube-shaped siphon like a snorkel: the siphon pokes up through the top front of the shell and reaches skyward through sand until it contacts open water containing fresh oxygen. Mating and spawning occur throughout the year. First, the male follows the ripe female's mucus tracks. Next, his foot touches her shell, and the two bond together in a sticky mass. Their union lasts for up to three days and is so tightly bound, they remain glued together even if rolled in the surf or touched. Predators include the armored sea star (p.41), octopuses (p.32, 33), and other snails.

Wart-Necked Piddock

Chaceia ovoidea

A piddock clam's head is reduced to just its mouth opening. The gills strain floating microscopic plants, then whisk them into its mouth. The siphons are light-sensitive and, when struck by a flashlight beam, quickly close down and contract into their sand burrow. The piddock's boring way is a major cause of canyon erosion and subsequent underwater landslides. Once a small slide occurs, the collapsed area often exposes the piddock's hidden shell and long, fleshy foot for all to see. The landslides are not a danger to divers.

Appearance: Two long, wrinkled siphons appear as smokestacks sprouting from the substrate. Siphons are whitish with brown flecks; siphon tips are deep mahogany-red. Exhalant siphon is bell-shaped; inhalant siphon is tube-shaped. Shell is white.

Size: Siphons to 2 inches in length. Shell to 4 inches in length.

Where Found: Along the canyon rim bored horizontally into shale ledges from 50-80 feet deep.

Feeds On: Microscopic plants.

Egg mass

Appearance: Body is usually bright yellow but may be warm red-brown; body is speckled with white dots. Gill tree resembles a white flower. Horns are club-shaped.

Size: To 1 inch in length.

Where Found: On canyon ledges from 55-85 feet deep, especially along North Wall.

Feeds On: Sponges.

Salted Dorid

Doriopsilla albopunctata

This shell-less snail has a muscular sucking capability that allows it to feed on soft-bodied animals. When the dorid approaches a sponge, for example, it ejects a small tube from its mouth and releases digestive fluids to soften the sponge; it then sucks the sponge into its mouth. Spawning occurs mostly during the summer when the dorid lays eggs in a tightly coiled yellow ribbon, which it fastens to a rock or algae. Upon sensing danger, the salted dorid retracts its "cottontail" gills into the small pocket on its back.

Adult with egg mass

Hermissenda

Hermissenda crassicornis

This aggressive sea slug fights frequently with other Hermissendas. When two engage in combat, they start by head-butting, which soon deteriorates into lunging and biting. The first one that inflicts a wound on the side or tail of its adversary usually wins the battle, and the loser slinks away in defeat. During the winter months, clusters of two to four animals can be found coupling in pairs or as a group; they attach their whitish egg strings (laid in narrow coils that resemble linked sausages) to a strand of algae or eel grass (p.13).

Appearance: Grayish-white translucent body is decorated with opalescent blue lines and two midbody gold lines; long gills (cerata) are brownish-yellow with white and gold tips.

Size: To 2 inches in length.

Where Found: On ledges or mud bottom in the canyon from 55-85 feet deep. Also sometimes found in eel grass (p.13) from 30-50 feet deep.

Feeds On: Small sea anemones, bryozoans, worms, dead animals, and various hydroids.

Appearance: Translucent and vivid blue-violet body; magenta horns; and bright orange gills (cerata).

Size: To 2 inches in length.

Where Found: Found along the canyon ledge from 50-85 feet deep.

Feeds On: Hydroids.

Spanish Shawl
Flabellinopsis iodinea

This gaudy sea slug swims as well as crawls. When it swims, it does so upside down. In this way, it reduces the resistance created by its protruding orange gills and, hence, becomes more aerodynamic. An artful predator, the shawl bites off the stinging cells of its prey and transfers the toxins to the tips of its own gills without firing off the poisons into its own body (it is not known how). While this self-preservation technique is impressive, its predator, the striped sea slug (p.31), is also unaffected by the poisons. During reproduction, the Spanish shawl lays salmon-pink, gelatin-type egg masses that cling to bits of seaweed.

Giant Dendronotus

Dendronotus iris

One of the largest nudibranchs, the giant Dendronotus reaches a walloping 11 inches elsewhere along the Pacific Coast. Despite its size, this Dendronotus can do more than just crawl along the bottom; by twisting its body from side-to-side, it can swim as well. During mating season, an adult lays long, loosely looped white egg strings; the larvae hatch in ten to twenty days.

Appearance: Pale pink body is covered with many branchlike protrusions.

Size: To 4 or 5 inches in length.

Where Found: Mostly on mud bottom deep in the canyon from 80-100 feet deep.

Feeds On: Tube-dwelling anemones (p.17).

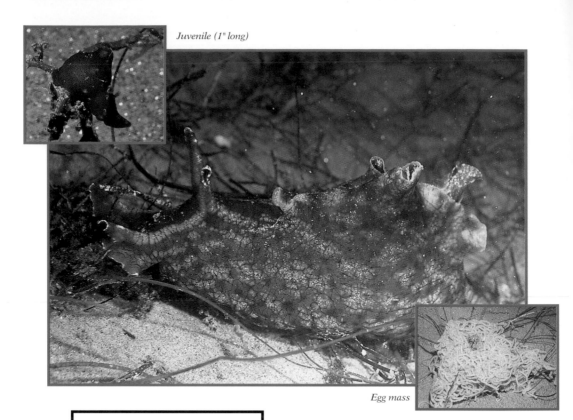

Juvenile (1" long)

Egg mass

Appearance: Flabby body is
mottled reddish-brown;
two "horns" adorn the head.
Resembles a crouching
rabbit. Juveniles (to 1 inch
in length) are solid brick-red.

Size: Typically 5-6 inches
but can reach over 1 foot
in length.

Where Found: Mostly in the
canyon, from 50-70 feet
deep but can be found
shallower.

Feeds On: Seaweed.

California Sea Hare
Aplysia californica

During the summer, orgies of three, four,
five or more sea hares entwine for several
hours or even days to produce piles of long,
tangled, yellow, yarnlike masses of eggs.
Throughout mating time, the sea hare serves
as either a male or female (a hermaphrodite).
It lays about 41,000 eggs per minute, and
each mass contains about 85 million eggs.
The eggs hatch in about twelve days, and
the precious few that survive live only one
or two years. As a protective mechanism,
this sea hare species squirts a thick cloud
of reddish-purple ink when disturbed. Still,
it cannot always avoid its predator, the
striped sea slug (p.31).

Mating orgy

Egg mass emerging

Striped Sea Slug

Navanax inermis

A predator, this shell-less snail sneaks up on its prey by tracking the victim's mucus trail, then swallows the mollusc – shell and all. Once the victim's soft body is completely digested, the only evidence of the silent attack is seen later when the sea slug eliminates the unbroken shell. After mating, the striped sea slug releases a string of yellow-white eggs in the shape of a fragile, loosely woven basket. Find the eggs attached to eel grass (p.13) or bits of algae.

Appearance: Cigar-shaped body is dark velvet-brown with a blue-violet sheen. Markings vary but usually include yellow or white streaks or spots; edges are trimmed with orange-yellow lines and spots of sapphire or turquoise blue.

Size: To 4 inches in length.

Where Found: Throughout the canyon, from 50-100 feet deep, on mud bottom or along ledges.

Feeds On: Various molluscs including California sea hares (p.30).

Appearance: Dull red, reddish-brown, to pink overall, often with white mottling. Tiny body is round to oval; arms are four times body length.

Size: To 2-4 inches in height including body and out-stretched arms.

Where Found: On the mud, in the canyon (often around seaweed) from 50-75 feet deep.

Feeds On: Crustaceans, molluscs, fishes, and small crabs.

Red Octopus
Octopus rubescens

In spite of its small size, this octopus is an efficient predator. Once it seizes a victim (a crab, for example), it kills the prey by secreting poison from its salivary glands. It then exposes the crab's stomach and eats the contents. Finally the legs are pulled off and cleaned out one by one. Nevertheless, this tiny terror has its share of enemies; it is preyed upon by various fishes such as the spotted sand bass (p.55), the kelp bass (Vol.1, p.40), and rockfishes such as the brown rockfish (p.52).

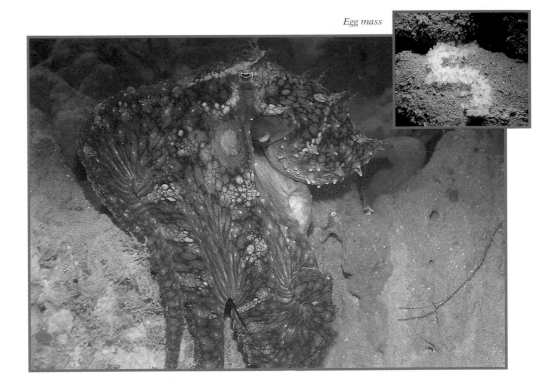

Egg mass

Two-Spotted Octopus

Octopus bimaculatus

Closely related to the market squid (pp.34-35), the octopus lays strings of white eggs, which the female broods for a couple of months until the eggs hatch. This nocturnal mollusc usually waits until dark to emerge and forage for food; during the day, it generally remains sequestered in the safety of its den. Therefore, even on a day dive, bring a dive light to shine into the many recessed holes you swim past. Sometimes an octopus may lean out of its hole to sneak a peak at you as you pass by. In this case, back off and drop down to hide from the octopus; the nosy cephalopod is likely to emerge to see where you went and what you are doing.

Appearance: Usually mottled-brown, but changes color to match its moods and background. Below its true eyes, two large, dark ovoid spots display a deep-blue color.

Size: To 3^{1}/$_{2}$ feet, including body and outstretched arms.

Where Found: Tucked into holes along the canyon's ledges; especially abundant around Vallecitos Point from 55-65 feet deep. Occasionally perched out in the open.

Feeds On: Crabs, California spiny lobsters (p.38), some molluscs, and small fishes.

Market Squid
Loligo opalescens

Appearance: Cylindrical body is translucent bluish-white, changing to mottled gold and brown or other colors and patterns when feeding, frightened, or excited. Huge iridescent eyes; short, broad arms attach to the top of its head.

Size: About 1 foot in length including arms.

Where Found: Anywhere along the canyon from 30-100 feet deep or more. Sightings vary from year to year; some years no squids are seen, some years uncountable numbers are seen.

Feeds On: Tiny shrimplike crustaceans, fishes, worms, and its own young.

Of all the ocean's phenomena, the return of the squid to spawn in La Jolla Canyon from November to March must be one of the most exciting and dramatic events that can be seen off the San Diego coast. Because it is a brief and unpredictable event, divers impatiently await this time of year in hope of witnessing the squid's elaborate mating ritual. When the squids arrive, they do so by the thousands. Over time, waves of squids appear, disappear, and appear once again as fresh squids arrive to take the places of ones that died. A squid lives most of its life (two years) in depths of at least several hundred feet and returns to shallow water (where it was born) to complete its life by spawning and then dying.

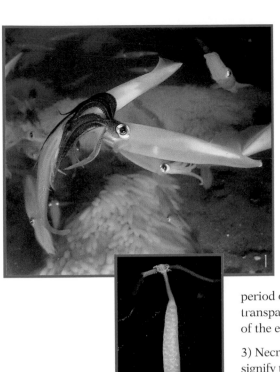

1) During mating, the male's arms blush red; he clasps the female and uses a specialized arm to transfer a sperm packet into her body cavity.

2) The cigar-shaped capsules, about the size of your index finger, contain 100-200 egg sacs that look like beads. Each female lays up to 24 egg cases over a period of a few days. She anchors her rubbery, transparent capsules by tying the stringlike part of the egg case to seaweed or sand grains.

3) Necrotic ulcerations on the arms and body signify the end of spawning and the end of the squid's life. Like salmon, once squid spawn, they die.

4) Millions of eggs carpet the ocean floor. The egg cases appear to have no predators, maybe because of their bad taste or toxic properties.

5) The red worm *Capitella ovincola* can be found in the egg cases, but not within the egg sacs themselves. It usually does not bother the developing squids because it feeds only on their waste products.

6) The squid takes about a month to develop, then uses its sharp, beaklike teeth to gnaw through its sac. About the size of your fingertip (1/5 inch long), the hatchling escapes at night to avoid predators, which include adult squids.

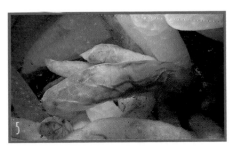

Appearance: Tiny, transparent, sticklike body has three legs and two arms; two long antennae protrude from the top of its head.

Size: To about 1 inch in length.

Where Found: Clinging to plants in Pipefish Patch (15-25 feet deep) and Eel Grass Island (30 feet deep), or in the canyon hooked to a fan in Gorgonian Gardens (80-100 feet deep).

Feeds On: Tiny crustaceans and decomposed plants and animals.

Skeleton Shrimp
Caprella californica

Anchoring its three legs to its post, the skeleton shrimp bows and scrapes, a behavior similar to that of a land-dwelling, praying mantis. It uses its two arms for cleaning, for defense, and to capture food. By using its legs and arms together, the skeleton shrimp moves in a quick inch-worm motion. When knocked off its perch, it swims in a jerky motion by rapidly flexing and straightening its body.

Red Rock Shrimp

Lysmata californica

Animals wanting to be groomed, including the California spiny lobster (p.38), visit the red rock shrimp's cleaning station. When a host drops by, the small shrimp hops onto its body, moves over its surface (even into its mouth!) and picks off almost anything removable. Divers find that this small shrimp will carefully clean their extended hand, especially the areas around the fingernails. The cleaning relationship, however, is not an exact science; aquarium studies have shown that a group of shrimp may clean a fish down to its bones. In the reverse, shrimp remains have been found in the stomachs of its known hosts as well.

Appearance: Thin white stripes traverse the back of its red body.

Size: To about 2 inches in length.

Where Found: In the canyon in cracks and crevices from 60-85 feet deep.

Feeds On: Parasites and other matter from the bodies of other animals, especially fishes.

California Spiny Lobster
Panulirus interruptus

Appearance: Red to orange coloration; spiny projections stud the carapace (upper shell) and the sides of its tail. The two antennae are twice as long as its body.

Size: To 2 feet in length along the carapace.

Where Found: On sand and mud bottoms, in crevices along the canyon's ledges, and under overhangs from 15-100 feet deep.

Feeds On: Plants and animals, living or decomposing.

To chase off other animals attracted to its prey or to frighten enemies away, the lobster deploys a sweeping motion with its extra-long antennae or broadcasts an alarming grating noise (from a structure at the base of each antenna) that can be heard even by divers. Mating occurs during winter and early spring when the male deposits a sperm pack on the underside of the female. While most notable for being good to eat, the lobster may also have a future in food preservation. One scientist created a preservative gel from the crustacean's chitin – the material that makes lobster shells hard – then used it to coat apples; the fruits stayed fresh for at least six months.

Sheep Crab

Loxorhynchus grandis

Because of its large size (probably the largest spider crab sport divers see), peculiar shape, and deliberate movements, the sheep crab presents a ludicrous appearance. As a juvenile, this timid crab hides in plain sight by attaching small animals and plants onto its body to match its background. As an adult, the sheep crab transforms from threatened to threatening as it walks with impunity out in the open. Should you decide to pick it up by the top of its shell, its arms will rotate around, and its huge, strong pincers will grab your fingers making you sorry you did! Mating occurs in spring and early summer.

Appearance: Inflated, oval, fuzzy-looking, greenish body with very long legs.

Size: To about 1 foot across carapace (upper shell).

Where Found: La Jolla Shores and canyon from 20-100 feet deep.

Feeds On: Living or dead sea stars (p.41; Vol.1, p.31), clams, octopuses, and market squid (pp.34-35).

Hemphill's Kelp Crab
Podochela hemphilli

Appearance: Delicate body is shaped like an isosceles triangle. First pair of walking legs are adorned with numerous curved hairs. Decorates itself with short pieces of brown or red algae.

Size: To about 1/2 inch across body.

Where Found: Along La Jolla Shores and in the canyon from 20-100 feet deep, clinging to plants or ledges.

Feeds On: Tiny animals living on kelp.

When frightened, this small kelp crab raises up one of its front legs and holds it horizontally between itself and its perceived threat. If a crab's leg is trapped or gripped by a predator, the cornered crustacean escapes by using special muscles that literally throw off the limb so that it can flee. Surprisingly, this auto-amputation technique barely affects the crab's mobility. A self-preservation trait as successful as this may be one reason that crabs have long been a part of history; they first appeared in fossil records nearly 200 million years ago.

Armored Sea Star

Astropecten armatus

Although the sea star wears a stiff suit of armor, it is surprisingly flexible. When flipped upside down by the surge, it bends its body back until it can flip the rest of itself over and return to its proper position. Divers take note of the many amputees (less than five-armed stars) regularly seen off La Jolla Shores. The sea star survives even the most brutal battles because of its wonderful powers of regeneration. In fact, somctimes the armored sea star heals itself so well, you may find it has regenerated a sixth(!) arm.

Appearance: Five-armed pink to lavender body is completely outlined by a row of white plates around its edges.

Size: To about 7 inches in diameter.

Where Found: On sandy bottom from 30-40 feet deep.

Feeds On: Snails, especially purple olives (p.24), sea pansies (p.15), and sand dollars (p.42).

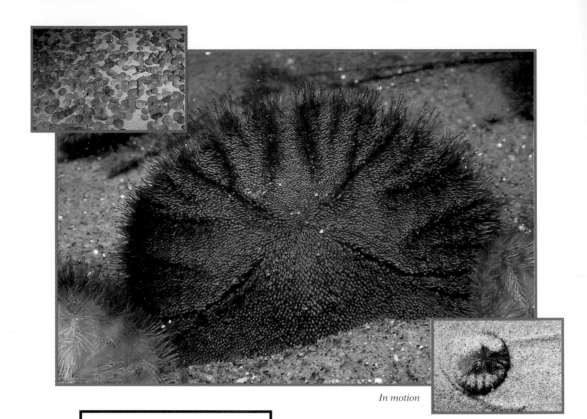

In motion

Sand Dollar

Dendraster excentricus

Depending on water conditions, a sand dollar aggregation can look like a military regiment at attention or a mish-mash of poker chips strewn on a table. In quiet waters with little surge and current, a sand dollar digs the lower edge of its body into the sand, while the upper part stands perpendicular to shore; in rougher water (outgoing tides), the animal chooses a lower profile by lying flat on the sand surface. An erect posture better permits the sand dollar to capture drifting food; a flat posture allows it to move, since it is the short, slender, mobile spines that drive the sand dollar's body forward across the sand. A sand dollar may live up to ten years. The California sheephead (p.56) is a predator.

Sea Porcupine

Lovenia cordiformis

Although it lives deep in the canyon for most of the year, the sea porcupine (also called a heart urchin) returns in large numbers to shallower water to mate. Its test (shell) is thinner and more fragile than other urchins. While the porcupine's extremely long spines usually lie flat against the body, these daggers can be erected and aimed at any source of irritation. Even though the spines are a good defense (they can inflict painful wounds if not handled with care), they are no match for bat rays (p.49), which find them easy prey; large sand depressions made by bat rays are often found filled with broken sea porcupine tests.

Appearance: Off-white, oval, spiny body tapers toward the rear; extra-long spines sprout from the area behind its eyes and point toward the back.

Size: About 1 1/2 inches in length and 1/2 inch in height.

Where Found: On sand and mud bottoms along La Jolla Shores and in the canyon from 25-100 feet deep.

Feeds On: Decomposed plants and animals.

Appearance: Gray body is adorned with thick black, elongated spots that drape over its back and sides; underbelly is white. Moves in a snakelike motion.

Size: To 7 feet in length, but usually 5 feet or less.

Where Found: On sandy bottom off the Marine Room in Shark City from 3-10 feet deep.

Feeds On: Clams, octopuses, crabs, California spiny lobsters (p.38), bony fishes, and bat rays (p.49).

Leopard Shark
Triakis semifasciata

The leopard shark is not dangerous and is, in fact, fearful of diver's bubbles or swimmer's movements. During the summer, large groups congregate on the shallow, sandy bottom to breed. After a twelve-month gestation period, females give birth to litters of from four to twenty-nine offspring, which are 8-9 inches in length. Juveniles grow up to 4 inches a year and have an average life span of about twenty-five years. In Southern California, fossil leopard sharks have been found in deposits over a million years old.

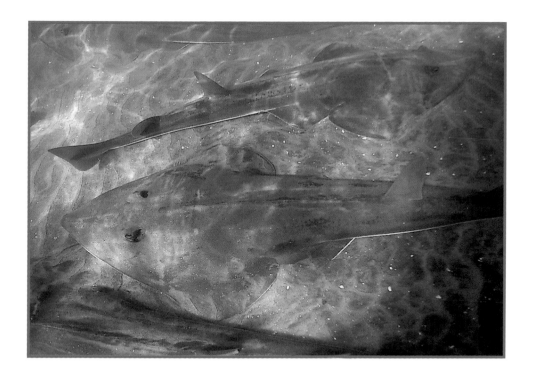

Shovelnose Guitarfish

Rhinobatos productus

On the evolutionary scale, the guitarfish
is one of the most primitive rays. Seen
sporadically year round, it is a regular
resident during the summer. It congregates
to mate in large numbers (fifty or more)
on the shallow sandy bottom in and just past
the surf zone. When feeding, it can become
so engrossed that it may strand itself when
the waters recede and be forced to wiggle
its way back into the ocean. Females bear
up to twenty-eight live young, which are
6 inches in length. The shovelnose guitarfish
lives at least ten years.

Appearance: Pale brown on
top with white underside;
flattened, spade-shaped
head, pointed nose, and
a long, thin tail; row of
spines align atop its back
and tail.

Size: To 5$\frac{1}{2}$ feet in length.

Where Found: La Jolla Shores,
especially in Shark City,
in depths from 3-30 feet
deep.

Feeds On: Crabs, worms,
shrimps, and fishes.

Thornback

Platyrhinoidis triseriata

Appearance: Heart-shaped, solid-brown body comes together in a rounded point toward the front of the eyes; three rows of large, hooked spines line the back and long tail.

Size: To 2¹/₂ feet in length.

Where Found: Off La Jolla Shores mostly from 5-30 feet deep; in the canyon to 100 feet deep. Often partially buried.

Feeds On: Worms, clams, crabs, and shrimps.

Although closely related to the shovelnose guitarfish (p.45), the thornback's spines easily distinguish it from any other ray. While used for defense, the thorns are magnets for various drifting seaweed looking for a place to anchor; a thornback ray often can be seen moving along the bottom with plants sprouting from its back. Mating occurs in late summer.

Buried with only gill slits exposed

Round Stingray

Urolophus halleri

A nonaggressive animal, this perceived demon can still cause grief to bathers. When inadvertently stepped on, the ray's tail bends over almost completely to reach its body (covering the hapless bather's foot). In doing so, it exposes a knifelike barb containing poison that it impales into the upper part of the foot (not the sole). To avoid the sting of its nasty barb, do the "Stingray Shuffle" (drag your feet along so as to not step on the ray) when entering the water. The shy ray will sense danger from the vibration and take off, leaving only a cloud of sand.

Appearance: Mottled-brown, gold, gray, or black coloration on its small, circular body. Short, thick tail harbors a long spine that reaches toward the tip of the tail.

Size: To 22 inches in length.

Where Found: Along La Jolla Shores in and just past the surf zone, usually completely buried, from 1-10 feet deep.

Feeds On: Worms, molluscs, shrimps, and crabs.

Appearance: Mottled-gray body is shaped like a butterfly with wings outstretched; tail is very short.

Size: Wingspan to 5 feet.

Where Found: La Jolla Shores from 5-30 feet deep.

Feeds On: Worms, crabs, shrimps, and molluscs.

California Butterfly Ray

Gymnura marmorata

This animal is grouped with other stingrays that include the round stingray (p.47) and bat ray (p.49); however, the butterfly's stinger is located at the base (not tip) of the tail and is so small that it is not particularly helpful for defense. The butterfly ray is somewhat difficult to find because of its pale color and habit of completely burying itself in the sand.

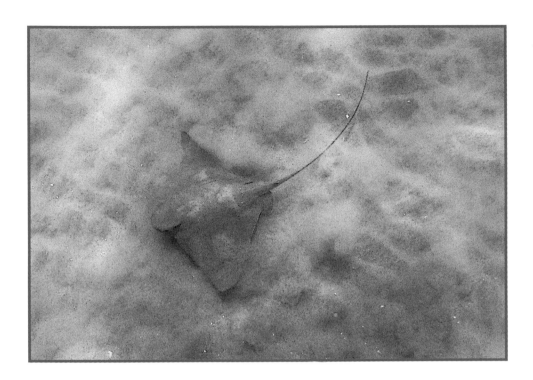

Bat Ray
Myliobatis californicus

This stingray has jaws of steel. With its impressive grinding teeth, an adult can exert a walloping 150 pounds of pressure through its mouth, which makes crushing a clam shell as challenging as chewing Jell-O. Adults give birth to live young. When born, a bat ray emerges tail first with its wings wrapped around its body like a double-rolled Mexican tortilla. Because a newborn's tail is rubbery, the mother is protected from being impaled by the baby's poisonous spine during the birth process; however, only a few days after birth, the spine hardens.

Appearance: Blackish or brown on top with white underside; head protrudes showing a distinct face. Base of its long, whiplike tail harbors a serrated stinger.

Size: Female to 6-foot wingspan; male to 2-foot wingspan.

Where Found: Usually off the Marine Room from 10-25 feet deep, but also seen in the canyon from 50-100 feet deep.

Feeds On: Oysters, clams, crabs, California spiny lobsters (p.38), shrimps, and fishes.

Bay Pipefish
Syngnathus leptorhynchu

Appearance: Wire-thin, bony, plated body; horselike head has long, narrow snout. Solid colored or mottled various browns, greens, or yellows to match algal background.

Size: To more than 10 inches in length.

Where Found: In the shallow waters of Pipefish Patch and occasionally Eel Grass Islands from 15-30 feet deep. Tiny juveniles, cling to surf grass (Vol.1, p.15) flotsam at the undersurface of the water.

Feeds On: Various tiny animals.

The pipefish closely resembles its cousin, the seahorse. Like a seahorse, the pipefish swims in a more or less upright position, sculling itself along, partly by using its fins and partly by wriggling its head, body, and tail. A pregnant pipefish is a male; he wears a brood pouch on the underside of his body to which the female transfers her eggs. They incubate for eight to ten days before hatching. This bony animal is easily able to change color and disappear into its weedy background, which may be the red alga thin dragon beard (p.12). Should you see a couple of floating strands of living or dead algal blades, look closely; one just might be a bay pipefish.

California Scorpionfish

Scorpaena guttata

Often called a sculpin, this fish looks quite lethargic as it sits motionless on a ledge. The scorpionfish can inflict a painful sting by imbedding its toxic spines in gullible victims, which include careless humans. Spawning takes place from April through August. The eggs, barely visible to the human eye, are enclosed in pear-shaped balloons that are 5 to 10 inches long. Once the balloons are released from the bottom of the sea, they quickly rise to the surface where they float protected until they hatch (within five days). The scorpionfish may live fifteen years or longer.

Appearance: Usually mottled-red and spotted but can change to ash-gray to match its background; thick-bodied with large spiny fins.

Size: To 17 inches in length.

Where Found: Resting quietly on mud bottom along the canyon rim or deeper into the canyon from 40-100 feet deep.

Feeds On: Fishes, market squid (pp.34-35), octopuses, shrimps, and crabs.

Appearance: Apricot with brown mottling. Wide, squat body has protruding spines. Immediately identifiable by dark-brown spot on gill cover.

Size: To 21½ inches in length.

Where Found: Most abundant along North Wall from 55-80 feet deep, either wedged in a crevice or perched on the ledge.

Feeds On: Fishes and crabs.

Brown Rockfish

Sebastes auriculatus

Brown rockfish may have a homing tendency. When several browns were captured in San Francisco Bay and released more than ten miles west of the Golden Gate Bridge, at least one returned home to where it was originally captured. Browns spawn more than once per year, and when they do, adult females may produce up to about 340,000 eggs. Because most rockfishes make for excellent eating, they are hunted relentlessly off U.S. and Canadian shores by sport and commercial fishermen. In the store, most rockfishes are often purposely misnamed snapper or perch to sound more appealing.

Egg mass

Painted Greenling

Oxylebius pictus

Its jailbird markings (one of the few fish on the Pacific Coast to have alternating bars on its body) are reason enough to give the painted greenling its other common name, "convict fish." The greenling is related to the rockfish (p.52; Vol.1, p.38) and scorpionfish (p.51). A female lays a golden egg mass in a nest on exposed rock surfaces. The egg mass, guarded by a male, contains as many as 2,200 eggs. The greenling is only active during the day. The first record of this handsome fish was made back in 1866. The painted greenling lives at least eight years.

Appearance: Long, pointed head; five bold, brown-red to brown bars on body; small, pointed mouth. Head has two pair of branchlike fleshy flaps.

Size: To 5 inches in length.

Where Found: Along canyon ledges camouflaged around cracks and crevices or hiding in plain sight from 50-80 feet deep.

Feeds On: Crabs and shrimps.

Male guarding eggs

Cabezon

Scorpaenichthys marmoratus

Appearance: Mottled-red males or mottled-green females, but either can change color to mimic background. Body is scaleless. Dorsal fin is prominent when raised. Head is large, and lips are full and fleshy. Branched, hornlike protrusions perch above its eyes, which are placed wide on its head.

Size: To 3 feet in length.

Where Found: Along the canyon's upper ledge, particularly around Vallecitos Point from 50-65 feet deep.

Feeds On: Shrimps, octopuses, crabs, abalones (e.g., Vol.1, p.20), and small fishes.

The monogamous cabezon returns to the same nesting place every year. Spawning takes place from March through November. The female lays up to 100,000 eggs over a period of days, and the male guards them until hatching. When first laid, the eggs are a fluorescent green but by hatching time they turn deep blue-purple (note the older blue eggs peeking out from below the younger green eggs in the above picture). Throughout the summer, a solitary male can be found perched atop an egg mass he aggressively guards. Predators include the Brandt's cormorant (Vol. 1, p.57).

Spotted Sand Bass

Paralabrax maculatofasciatus

Most spotted sand bass start out as females, then switch sex to become males. The smallest male found was only 4¹/2 inches in length, so sex change can occur early in a bass's life. It spawns in late spring and early summer. A good-eating fish, the bass is popular with party-boat anglers.

Appearance: Gray body is overlaid with black spots; mouth is large with a protruding jaw. Juvenile bass also adorned with thick, black vertical bars.

Size: To 1¹/2 feet in length.

Where Found: Perched just above the canyon rim on sandy bottom and around Eel Grass Islands from about 30-40 feet deep.

Feeds On: Crabs, shrimps, and small fishes.

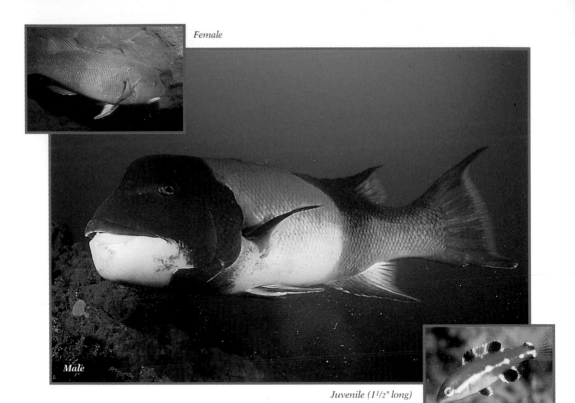

Female

Male

Juvenile (1¹/₂" long)

California Sheephead

Semicossyphus pulcher

Appearance: Male has white chin, black head, red or pink middle, and black tail; prominent bump defines forehead. Female is pinkish overall with white chin. Adults have imposing canine teeth. Juvenile is salmon-colored with white horizontal stripe across sides of body; fins have black dots.

Size: Male to 3 feet; female to 2 feet; juvenile to 4 inches in length.

Where Found: Above or along the canyon's upper ledge from 50-70 feet deep.

Feeds On: Sand dollars (p.42), clams, abalones (e.g., Vol.1, p.20), crabs, and California spiny lobsters (p.38).

Beginning life as a female, the sheephead changes sex after about eight years of age when its sex organs metamorphose into male sex organs. Sex transformation takes less than a year. Although each male has his own territory, should he be eliminated, a resident female will quickly fill the gap by becoming a male. The sheephead seems to prefer hanging around the outermost outcroppings such as the promontory at Vallecitos Point. These fish are curious about divers who enter their domain. Expect the large nosy males to hang around and watch what you are doing.

Sarcastic Fringehead

Neoclinus blanchardi

This bad-tempered, big-mouthed, aggressive fish charges and snaps at anything that passes its burrow. The fringehead will viciously attack any nosy diver by using its cavernous mouth to grab the closest body part it can. Fortunately for the diver, this show of rage is really more bark than bite, since the fringehead's giant jaws lack substantial teeth. The fringehead is found in the same family as the giant kelpfish (Vol.1, p.54).

Appearance: Large head tapers to a tail-like body; treelike branches protrude over eyes. Body is mottled-brown with two blue spots adorning its dorsal (back) fin; the mouth is outlined a cadmium yellow.

Size: To 1 foot in length.

Where Found: Usually burrowing in mud bottom (but may be found hidden in a shell) in deeper canyon depths from 80-100 feet deep; prominent in Gorgonian Gardens.

Feeds On: Clams, other molluscs, and small fishes.

Appearance: Orange to red, but changes color to match its surroundings; color uniform, mottled, blotched, or barred. Fleshy flaps (three pair) that look like tree branches protrude out upper eye edges. Yellow tail fin. Cone-shaped teeth.

Size: To about 2¹/₂ inches in length.

Where Found: In the canyon along ledge sandstone out-croppings from 50-80 feet deep, usually holed up with only its head exposed.

Feeds On: Tiny shrimps, molluscs, and fishes.

Yellowfin Fringehead

Neoclinus stephensae

Don't be fooled by the yellowfin's diminutive size. With an enormous mouth that extends far back behind its eyes – and rows of pointed teeth – this little fringehead is as fearless and aggressive as its relative, the much larger sarcastic fringehead (p.57). When approached by a diver, it may retreat for a moment, but its nosy nature and need to protect its territory compel it to come back out to see what is going on. The yellowfin is also in the same family as the giant kelp-fish (Vol. 1, p.54).

Blue-Banded Goby

Lythrypnus dalli

A tiny territorial beast, the brilliantly colored blue-banded goby is found boldly perched on exposed rocks but retreats in less than a blink of an eye when threatened. Goby species are among the smallest fishes in the sea. The goby is most abundant in the tropics where it hides around coral reefs. It is estimated that there are 2,000 species worldwide.

Appearance: Slim, oblong, coral-red body overlaid with from four to nine electric-blue vertical bars.

Size: To 2 1/2 inches in length.

Where Found: Steep slopes in cracks and crevices along the canyon wall from 50-80 feet deep.

Feeds On: Algae and tiny crustaceans.

Appearance: Large black spots adorn a flat, gray body. Both eyes usually lie on its left side. Mouth is filled with sharp, pointed teeth.

Size: Female to 5 feet; male to 2 feet in length.

Where Found: Generally at least partially buried in sand bottom from 20-80 feet deep.

Feeds On: Fishes, market squid (pp.34-35), and sometimes octopuses.

California Halibut
Paralichthys californicus

Although the halibut is born with an eye on each side of its head, one eye slowly migrates so that by adulthood both eyes are found only on one side. In this species, slightly more than half are left eyed. An energetic predator, this flatfish has been seen jumping clear out of the water as it makes passes through small fish schooling near the surface. The largest California halibut recorded was 5 feet long and 72 pounds. Spawning takes place from April through July. Predators include sharks, such as the leopard shark (p.44), sea lions (Vol.1, p.58), harbor seals (Vol.1, p.59), and fishing enthusiasts.

Sightings Checklist

MARINE PLANTS

____ Eel Grass

____ Thin Dragon Beard

INVERTEBRATES

____ Armored Sea Star

____ California Sea Hare

____ California Spiny Lobster

____ Chestnut Cowrie

____ Giant Dendronotus

____ Hemphill's Kelp Crab

____ Hermissenda

____ Lacy Bryozoan

____ Lewis' Moon Snail

____ Market Squid

____ Purple Olive

____ Red Gorgonian

____ Red Octopus

____ Red Rock Shrimp

____ Salted Dorid

____ Sand Dollar

____ Sea Pansy

____ Sea Porcupine

____ Sheep Crab

____ Simnea

____ Skeleton Shrimp

____ Slender Sea Pen

____ Spanish Shawl

____ Strawberry Anemone

____ Striped Sea Slug

____ Trough-Nose Worm

____ Tube-Dwelling Anemone

____ Wart-Necked Piddock

VERTEBRATES-FISH

____ Bat Ray

____ Bay Pipefish

____ Blue-Banded Goby

____ Brown Rockfish

____ Cabezon

____ California Butterfly Ray

____ California Halibut

____ California Scorpionfish

____ California Sheephead

____ Leopard Shark

____ Painted Greenling

____ Round Stingray

____ Sarcastic Fringehead

____ Shovelnose Guitarfish

____ Spotted Sand Bass

____ Thornback

____ Yellowfin Fringehead

Notes

Bibliography

Dawson, E.Y. (1966). *Seashore Plants of Southern California.* University of California Press, Berkeley and Los Angeles.

Garfield, J.L. (1994). *The San Diego-La Jolla Underwater Park Ecological Reserve: La Jolla Cove, Vol. 1.* Picaro Publishing, San Diego.

Goodson, G. (1988). *Fishes of the Pacific Coast.* Stanford University Press, Stanford.

Gotshall, D.W. (1989). *Pacific Coast Inshore Fishes.* Third Edition. Sea Challengers, Monterey.

Love, R.M. (1991). *Probably More Than You Want to Know About the Fishes of the Pacific Coast.* Really Big Press, Santa Barbara.

Miller, J.M. & R.N. Lea (1972). *Guide to the Coastal Marine Fishes of California.* Fish Bulletin No. 157. California Department of Fish and Game, Sacramento.

Morris, R.H., D.P. Abbot, & E.C. Haderlie (1980). *Intertidal Invertebrates of California.* Stanford University Press, Stanford.

Moratto, M.J. (1984). *California Archaeology.* Academic Press, Inc., Orlando.

North, W.J. (1976). *Underwater California.* University of California Press, Berkeley.

Schaelchlin, P.A. (1988). *La Jolla, The Story of a Community 1887-1987.* Friends of the La Jolla Library, La Jolla.

Index